Way Up North Where the Kittiwakes Play

An A to Z Alphabet Book for Child and Adult

Juli-Ann Gasper

Photography by Joseph Gasper
and Juli-Ann Gasper
Omaha, NE USA

Printed in the United States of America

ISBN 979-8-89114-009-7 (sc)
ISBN 979-8-89114-010-3 (hc)
ISBN 979-8-89114-011-0 (e)

Library of Congress Control Number: 2023915911

2024.07.09

MainSpring Books
5901 W. Century Blvd
Suite 750
Los Angeles, CA, US, 90045

www.mainspringbooks.com

Table of Contents

Big Idea: Intellectual Property!

Why is it important to acknowledge the work of another person when you are writing something? In the Photography Acknowledgements list, what does "public domain" mean? What do you think a "commons license" is? If you want to cite THIS book in something you write, here is the correct citation:

Gasper, Juli-Ann. *Way Up North Where the Kittiwakes Play*. Omaha, NE. 2023.

Walrus Haul-Out

Preface

The stories and pictures for this alphabet book come from an expedition to Svalbard Archipelago, Norway, in July 2012. The expedition was sponsored by the National Geographic Society and operated by Lindblad Expeditions using the ship *National Geographic Explorer*, a fully-stabilized, ice-class vessel, with an Ice-1A rating on the hull.

All photography, unless otherwise noted, was by Joseph or Juli-Ann Gasper. Information about the topics is from seminars and workshops during the expedition, as well as a few outside sources of information (available on request to the author). For further information, contact Juli-Ann at jgasper2@iCloud.com.

How to Use This Book

As a picture book—point, talk, identify colors, count, pretend to be a bird, a bear, or a walrus

Big Words—for children and adults who love big words

Additional information—each alphabet page will have an additional information page. You can tell your child this information in your own words, or not! As you choose!

Vocabulary List—the word list index shows every alphabet letter where you will find information relating to that term.

A

Arctic Circle—What is really cool about the Arctic Circle is that the sun never sets north of this line, at least one day a year. For the island archipelago Svalbard, the sun never sets from mid-April until mid-August! It just goes around in a circle in the sky!

Alaska   North Pole

Arctic Circle map

Svalbard Archipelago

Big Word: Archipelago!

Additional Information:

The Arctic Circle sits at 66 degrees north of the equator in 2012 (it actually moves about 15 meters a year, based on the earth's tilt! I bet you didn't know that!)

An Archipelago is a group of islands. They are often volcanic, but can also result from erosion, deposition, and land elevation. Svalbard Archipelago is at 74 to 81 degrees north, about 1300 miles north of Oslo, Norway. 60% of the islands are covered with glacier ice or ice cap. Sea ice exists all year round in the most northern parts of the archipelago. There are only 2600 residents in Svalbard, 2100 of whom live in Longyearbyen (pronounced "long year bee un"), which has the northernmost-in-the-world school, church, hospital, bank, airport, hotel, newspaper, movie theater, and university. The main industries in Svalbard are mining (about 400 people), scientific research, tourism (3 hotels and an active art community) and education (including a university with 400+ students) and support services (government, safety, retail, shipping) for the 2100 residents. Svalbard does not require a visa or work permit or other documentation to live there. You must be able to support yourself. If you can't, or become ill, or cause problems with others, you will be "returned" to the place from which you came. Basically, it is a "no tolerance for mischief" zone! The archipelago is nominally governed by Norway, but 46 nations have signed a treaty for scientific and cultural equality and non-discrimination in the zone. The main ethnic groups are Norwegian, Russian, Philippine, and Thai.

B
Bear   Ice Bear   Polar Bear   Sea Bear

How many toes does this bear have? This is a polar bear, also called an ice bear. She lives on the sea ice and waits patiently at a seal air hole until one comes up for a breath. Then she grabs it! In the picture here, the bear has

What do you think the "bear in a triangle" sign means?

captured a baby beluga whale and is defending it from other nearby bears that can smell the food.

Big Word: Barnacles!

Additional Information:

Barnacles are crustaceans, related to lobsters and crabs. They attach themselves to hard surfaces, such as boat sides and whale tails! This means that they are mobile, traveling wherever the hard surface goes! Check out the barnacles on the whale's tail on the "W" page.

Ice Bear: Also called Polar Bear or Sea Bear. The bear on the upper left is a female, relatively healthy. She strolled right up to the bow of the ship and "explored" us. There are strict noise rules on board. When wildlife is near, there is no talking allowed; doors may not be slammed; the engines are turned off. The bear in the lower right hand picture is an old, malnourished male. We could see his vertebrae when he was lying down. We watched for at least 45 minutes while he fought off challenge after challenge. There were at least four other bears trying to get to the food, but he was clearly VERY HUNGRY. One adult who tried was a mother bear with one cub. She was malnourished and had probably lost her second cub already. So she really needed the food.

The bear sign, upper left, is posted at the outskirts of Longyearbyen, to warn folks that bears do come to the "city". All dogs have to be kept kenneled in a city-operated kennel and are allowed out only for sledding practice (on wheels in the summer) or on leash with the owner. No cats are allowed in Svalbard ("bear bait")!

C

Calve (a verb, action word) **Calf** (a noun, thing)

How many birds are sitting on the iceberg that **"calved"** off this ice cap?

The mother walrus is facing away from us; the baby, called a **"calf"**, is riding on her back! See how short her tusks are? Do your teeth grow longer when you get older?

Big Word: Pinniped!

Additional Information:

Icebergs calve off the seaward edge of a glacier or ice cap. In July 2012, just after we returned from our expedition, the news reported that an iceberg the size of Manhattan (46 square miles!) had calved off of an ice shelf in Greenland during the weekend. There has been a dramatic decline in arctic ice in the last 3 decades. This graph shows data for 1979-to 2023.

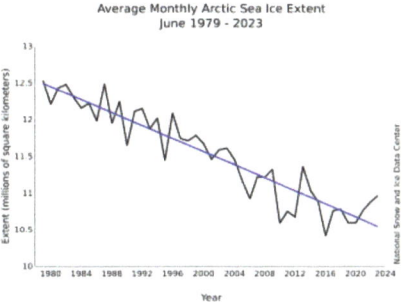

A baby walrus is called a calf. The walrus is actually a member of the seal or pinniped family! The species name is Odobenus rosmarus. It is the largest marine mammal other than whales. A large dense patch of up to 700 whiskers lines the snout. These are used to search for prey in the muddy sea bottom. The naturalists told us that each whisker is individually controllable by the walrus brain! They primarily eat clams by sucking the clam right out of its shell. Walrus cows (the mothers) take care of one calf at a time, for up to two years. The calf weighs about 150 pounds when born, but adult males grow to nearly 4,000 pounds. This cow and calf came within arm's length of our zodiac (see the Z page!). She seemed to be carrying the calf on her back.

D

Decontamination Dolphins

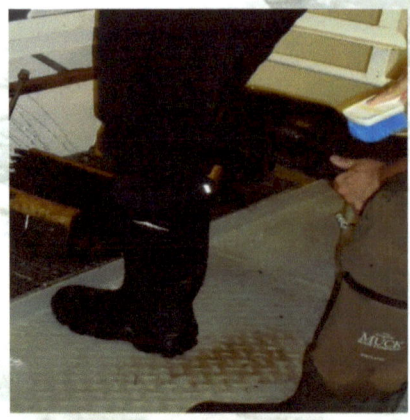

Why do you think the man is cleaning the boots so well? Aren't boots supposed to be muddy and dirty? This cleaning happened BEFORE we even went out on land and again when we came back to the ship. Once the boots were sanitized we had to leave them in a special room! Why?

Dolphins

Big Words: Permafrost! Decontamination!

Additional Information:

Decontamination of outerwear (parkas, hats, gloves, waterproof pants, boots, backpacks) ensured that none of our gear would carry onto the tundra any foreign seeds or critters. Each disembarkation from the ship followed the same procedure—a step into the sanitizing bath. Return to ship was a little more involved because we were often in mud, so the boots and pants had to be washed clean. Why was there so much mud? Because of the Permafrost! The ground freezes solid in the winter and then begins melting from the top down during the summer. The summer is so short that it only thaws about 3 feet down. All the water that was in the ground can't seep down (because it is frozen underneath!) so it stays in the first layers of dirt, thus creating mud. On one expedition to shore, four travelers sank into the mud from knee to waist deep. It took considerable time to get them out. One boot was not recoverable, so unfortunately the Arctic has a souvenir from our trip!

White-beaked Dolphins live in the north Atlantic. Their population, breeding habits, and life expectancy are mostly unknown. They live in deep waters most of the year. We were off of the continental shelf over deep water when we found these dolphins.

Explorers

E What will these explorers find from their Zodiac boat? How far do you think they could travel in it? How long could they travel in it?

What do you think she will find down so close to the ground?

What do you think she will find walking around? What can you see in the picture that she might explore?

Big Word: Exploration!

Additional Information:

An expedition like this one practically forces you to be an explorer! The plants on the tundra are so tiny that it is hard to see them from a standing position. We were astounded one day to find a grass clump that was all of two inches tall. It was the tallest vegetation we had seen for many days. This girl in the photo was the youngest on the trip. She continually explored the miniature gardens that grew on the south side of tiny slopes. Even a 4 to 6 inch rise is enough to protect arctic flowering plants from the wind and enable them to flower. The purple saxifrage (see the "X" page) blooms just days after the snow melts. So the naturalists can tell almost exactly when the snow melt occurred.

When you are on the shore or in the zodiac boat you can begin to imagine the horrific conditions that face those who live or have explored in this environment. We were wearing parkas, hats, and gloves in the middle of summer! The average winter temperature hovers around 0 degrees Fahrenheit. Do you know how to convert that to Celsius, the temperature scaled used in most of the rest of the world except USA and a few other countries? Take the F temp, subtract 32, then multiply by 5/9 and you have the C temp! So, for example: 0 degrees F - 32 = -32 * 5/9 = -18 degrees C.

F Can you find her in this picture? What season do you think is shown in this photo?

Arctic Fox

What season is this? What makes you think that?

What season is this? Why did you answer the way you did?

Big Words: Camouflage! Countercurrent Heat Exchanger!

Additional Information:

The arctic fox lives in the most extreme frigid parts of the Arctic North. It has several adaptations for cold survival, including very deep thick fur, high supply of body fat, a stubby body to reduce surface area exposed to cold, and short, thick ears. Also the fox has a countercurrent heat exchanger system in the blood vessels of its legs to keep its body heat concentrated in the body. Also, you can see in the picture how the arctic fox curls itself up and uses its tail as a blanket! Its fur is brown in the summer and whitish in the winter to blend in with the snow. In springtime, the fur changes gradually from white to summer brown and then back again in the fall.

The arctic fox has such keen hearing that it can hear prey under the snow and ice, like voles, lemmings, and ring seal pups. When it finds prey, it pounces on the spot and digs down to catch its meal. It also loves bird eggs and often has a den under a bird nesting cliff, like the one shown in the far left picture. This fox's den was at the base of a guillemot "cliff dwelling" where it could capture eggs that rolled off the ledges. (See the next page "G" for more information about these eggs!) Notice how hard it is to see the fox because its brown summer coat is perfect camouflage with the arctic tundra!

G

Glacier There are two glaciers here! What do you think happens when two glaciers bump into each other? What is all the black stuff?

Guillemots. These sea birds nest on cliff ledges.

Each female lays a single egg, and each is a work of art! Aren't they lovely? Why do you think the eggs are so pointed?

Big Word: Guillemot!

Additional Information:

Guillemot (pronounced "gill uh maht") This is a sea bird that lives on vertical cliffs along the shore. Hundreds of thousands will occupy one cliff, as seen in the middle picture.

They lay their eggs right on the cliff ledges. The eggs are larger than a goose egg and are very pointed at one end and fat at the other. The developing chick attaches to the fat end, making that end very heavy.' So when all the jostling of the adults moves the egg, it twirls in a circle, rather than rolling off the cliff!

Another interesting adaptation is a springy curl of feathers under the breast. The curl is about 3 inches wide and each feather is two inches long and is in a tight curl, about 1 inch in diameter, like a tube of feathers. To see what it might look like, curl your fingers into your palm. If the fingers were all feathers, you can imagine how the curl or tube would bounce. When the chick falls off the cliff after hatching, it bounces like a tennis ball! Very funny looking. The feather spring also acts as a shock absorber to protect the lungs of the adult from collapsing during deep sea diving. They will go as deep as 60 to 100 meters to feed on fish and crustaceans.

H

Haul-Out

Hump-Back Whale

Did you know that animals have "fingerprints" too? If they don't have fingers, what could be used to tell one from the other? Look at the walruses in their "haul-out". Can you tell them apart? What could be the "fingerprint" of a humpbacked whale? Do you think there is a "tail-print" database somewhere in the world? What would scientists use it for?

Big Word: Haul-Out!

Not really a BIG word, but certainly a fun one!

Additional Information:

Whale tail prints: The underside of each whale's tail is unique. This one had an underside that was almost completely white, but with very distinctive black splotches. These pictures will be submitted to the whale tale registry. If this whale has been previously registered, there will now be a record of one of its travel destinations! If not, then it will be a new submittal and its life in the database of whales will begin! We saw another whale whose underside tail was nearly completely black. Explore whale tail prints at www.happywhale.com

Haul-Out: The walruses pile onto each other to stay warm. During the summer they are building a 6 inch thick layer of blubber to keep warm in the winter. But summer temps are not very warm either, and trying to keep warm uses up energy that could be used to build blubber. So keeping warm makes them more efficient in the blubber-building process. The walrus also has up to 1 inch of "leather" skin to protect the blubber! It is so thick it can be sliced into layers for use by native peoples for shoes, harnesses, hats, etc.

I

This is a picture of the third largest icecap in the world. It is at least 100 feet high, so it must weigh a LOT! If you were wearing an ice-cap on your head that was the same height as *The Cat in the Hat* wears, would it push you down? What would happen as it melted away? Would you feel taller without all that weight holding you down?

Big Word: Isostatic Rebound!

Additional Information:

The weight of an ice-cap is tremendous. One cubic foot of ice weighs 57 pounds. So if you were wearing a 6 inch ice-cap, it would weigh about 30 pounds, and might bend you over with the weight! As it melted, you'd gradually be able to spring back up until you reached normal height again. So think about the weight of an ice-cap that is 100 feet high. WHOA! That would be 57 * 100, almost 6,000 pounds per square foot of surface! As an ice-cap melts, the earth underneath that has been compressed for all those thousands of years is able to spring back up. This is called isostatic rebound. So the land that is now under an ice-cap will be at higher elevation when the ice-cap is gone, yet another effect of global warming. This site has a nice video explaining Isostatic Rebound.

Of course, the amount of melting is offset by the amount of new snowfall, so it is a complicated equation to predict the effects on ice-caps. In some places, global warming causes HEAVIER snowfall, so the ice cap might increase in size, rather than decrease.

J Parasitic Jaeger

The Parasitic Jaeger steals food from other animals.

What is the name of the other animal in this picture? What do you think it is carrying?

Big Word: Kleptoparasite!

Additional Information:

This bird is called the Parasitic Jaeger (pronounced "jay grrr") because it pirates food from other animals! It is a kleptoparasite! It is also called a skua (pronounced "sku uh", with the same beginning sound as in "school"). It nests on the dry tundra of the arctic, and has olive green/brown eggs, as many as four at a time, a high number for arctic birds. It will dive-bomb the head of any animal approaching its nest, even a human. It feeds on rodents, and small birds, but loves to steal food from gulls and terns.

Other arctic animals are also kleptoparasitic. Gulls that can't dive will steal the fish that diving birds bring to the surface. Some species are kleptoparasitic as their major source of food. Others are kleptoparasitic from desperation of hunger. Ice bears that are stuck on land because their sea ice has disappeared may pirate the food of foxes and sea birds.

K

Kayak Kittiwake

Which is a **kittiwake**? A **kayak**? A dolphin? Does "dolphin" begin with "K"? Why are the kittiwakes following the dolphins?

Big Word: Onomatopoeic!

Additional Information:

Kittiwake is a seagull. During breeding season, the male and female nest on cliffs, producing 2 to 3 eggs. Pairs that have bred before tend to divide the egg/chick tending pretty equally. The Kittiwake has a yellow bill and black legs. These gulls can live 20 to 25 years. They are surface feeders. This explains why they are following the dolphins in the lower photo. The dolphins have stirred things up, making small fish more easily available for the kittiwakes. The call of the kittiwake is onomatopoeic (pronounced "ah no mah toe pea ic") If you say it like "kitti waaaaaake", it sounds sort of like their call. Sometimes we could hear the resemblance, sometimes not, so maybe they have different calls used for different purposes! You could have fun with your child listening to different bird calls and trying to imitate them. Some are pretty easy to mimic! "Chick-a-dee-dee-dee-dee". Bird books show the notes for bird calls so you can get the pitch right.

Sea kayaks are more like a wide, well-balanced, open top canoe than the kayaks we see on river rapids trips that cocoon your whole lower body. We had only one day when the sea was calm enough for kayaking. You can see in the picture how calm the water was. If you wanted to kayak, you had to go to a kayak training seminar. Here is a great place to talk about boating safety and life jackets!

L

In **Longyearbyen**, house colors are assigned by the Chief Color Chooser in Bergen, Norway. Pretend that you are the Chief Color Chooser. Can you find each of the house colors somewhere else in the picture or in the **lichen** picture?

Big Word: Symbiotic!

Additional Information:

In Longyearbyen, say you need to build a structure. The Chief Color Chooser (or probably some more official sounding title!) designates the color for every structure. The key consideration is that the colors must reflect the colors of Svalbard. What a lovely concept! There is another requirement: If something on the outside of a building was there before the establishment of the Community Council, it has to stay. A man from Australia put a kangaroo sign on his house. It is still there even though he is long gone; they can't take it off.

Lichen (pronounced "like un") is a symbiotic mix of algae and fungi that live together, forming a new plant. The algae provide the food through photosynthesis and the fungi retains moisture to avoid dehydration in the extremely dry Arctic air. This combination allows the lichen to live directly on a rocky surface where nothing else can live. As the lichen dies and decomposes, soil formation begins. Mosses find a foothold. Centuries later, there may be enough soil for microscopic plants to take hold.

As an ice cap melts, the land springs upward (isostatic rebound). As you walk from the shoreline up the beach to higher levels, you can identify the length of time since that ice cap melted by seeing the changes in the plant life from the first lichen to the mosses to the tiny plants.

M Mining Moss

Coal mining is a major industry in Svalbard. Do you know what coal is used for?

Do you know how coal is made by the earth?

How long does it take to make coal before it can be mined?

Big Words: Methane! Milankovitch!

Additional Information:

Methane, the most abundant organic compound on earth, is considered a greenhouse gas; it is manufactured by the earth by the biodegrading of organic materials. This happens currently (for example, from livestock belches and farts, and from rice fields and landfills) and anciently (trapped methane under sea ice, under ocean floors, releases from peat moss in permafrost thawing, and in coal deposits). Methane has a larger effect on greenhouse gases than carbon (25 times) but a shorter time in the atmosphere (8 years versus 100 years) before being broken down. Many industries work to capture methane being released by their work and use it as fuel.

The Milankovitch cycle is one possible explanation for naturally occurring warm/cold cycles over thousands of years. Milankovitch, a Serbian engineer and meteorologist, proposed that the solar energy received by the earth has three confounding factors—a wobble in the axis (like a spinning top wobbles), repeating changes in the shape of the orbit between circular and elliptical, and change in the tilt of the orbit. If you put these together, a cycle of warm and cold occurs naturally over 100-400 thousand years. So this cycle would need to be taken into account in assessing effects of human activity on global warming.

N

What do you think a **naturalist** does for a living? Do you think they have to dress in purple? Is the little girl a naturalist, too?

Big Word: Nitric Oxide!

Additional Information:

Nitric oxide is another of the greenhouse gases. It comes from nitrous oxide, which many people call "laughing gas", used as an anesthetic. Nitric oxide reacts with ozone and is the main naturally occurring regulator of stratospheric ozone (which is about 5 to 30 miles above earth's surface). It is considered an air pollutant with almost 300 times the impact of carbon dioxide (when considered over 100 years' time). Why do we care about ozone? See the "O" page!

A naturalist is a scientist who helps people gain appreciation for the natural world. Naturalists are focused on teaching people how sustainable ecosystems can be maintained or established. Naturalists are important team members who develop nature and cultural programs. A college degree in biology, geology, environmental studies, horticulture, or natural history will prepare you for an entry level position as a naturalist. There is growing demand for naturalists, estimated by the Bureau of Labor Statistics to grow by 12 percent in the next decade. Jobs may pay anywhere from $35 to $75 thousand annually. And think of all the wonderful places you will get to go!

O

Oystercatcher

Oyster

Mussel

Limpet

Crab

What do you think oystercatchers eat? NOPE!!!! Not oysters! It is really hard for a bird to open an oyster! They like mussels, limpets, worms, and crabs.

Big Words: Ozone!
Indicator Species!

Additional Information:

Oystercatchers are sea birds. They have huge beaks with very broad tips that they use to pry open mussels or hammer them on a rock to get the meat from the shell. Individual birds specialize in one technique or the other which they learn from their parents! Because of large numbers and readily identified behavior, scientists use the number of oystercatchers as an important indicator about the health of the ecosystems where they live. Arctic plant indicator species include lichen (see "L" page) and purple saxifrage (see "X" page). Reindeer are the most important arctic animal indicator species (see "R" page).

Ozone is sort of bi-polar! Near ground level it causes damage to tissue of plants and animals and so is considered a pollutant. But in the stratosphere it protects both plants and animals from ultraviolet light.

P

Ptarmigan Svalbard Poppy

The Rock Ptarmigan changes colors from winter to summer. This is called "molting". Why do you think this female looks so incredibly messy?

Big Word: Ptarmigan!

Additional Information:

Ptarmigan is not really that big of a word, but it is a very COOL word because it has a silent P at the beginning. It is pronounced "tar mi gun"! Some other names for ptarmigan are grouse and partridge. Our naturalist told us that in a big winter snowstorm, the ptarmigan will hurl itself into a snow drift and then stay there until the storm is over! Weird, huh?

When winter is over, the ptarmigan molts its winter white and changes to browns and grays to camouflage with the tundra. That's what is going on in the picture with all the shaggy feathers.

It is primarily a vegetarian, eating seeds, sprouts, young leaves, flowers, berries. If there are any insects in its area, the young feed on them.

What are some other words that have silent letters?

Q Quick! What color are bird beaks?

Ivory Gull

Puffin

Oyster-catcher

Arctic Tern

Barnacle Goose

Northern Fulmar

Big Word: Geolocator!

Additional Information:

Ivory gulls might be the most beautiful scavengers in the world, with pure white plumage and black legs and feet. They live almost exclusively off meat from polar bear kills, so their territory is almost completely in the Arctic North.

Puffins are sometime called the "penguins of the north" because they resemble penguins in coloring and stance. However, penguins live only in the Southern Hemisphere! No penguins at the North Pole! Puffins dive for their food, mostly small fish. They have short wings so they can "fly" underwater. They shed the bright colored beak after the breeding season!

The Arctic Tern has the longest migration of any known animal. It travels from North Pole to South Pole, a round trip of 44,300 miles, and gets to have two summers, no winter, and more daylight than any other creature on the planet. They can live as long as 30 years, so that means they travel as far as three trips to the moon! A geolocator, a very small device attached to the bird's leg, measures light intensity twice a day. From that, the scientists can determine where it is. The tern has dark red legs and feet to match its bright red beak! They mate for life and return to the same nesting grounds each year.

Rifle Reindeer

Why do you think our naturalist carried a rifle every time we were on land? What is the reindeer doing? Why did he lose the clump of hair that is underneath the letter R?

Big Word: Omnivorous!

Additional Information:

Svalbard is home to a distinct population of reindeer; small, with short legs, snubbed muzzles, and chubby bodies. Reindeer vary considerably in size, shape, and behavior from place to place and herd to herd. There are no natural predators in Svalbard. Reindeer fur is made of hollow hairs, with air inside, a great insulator from the winter cold. In the spring, the white winter hair comes out in clumps and brown hair replaces it. Ice bears also have hollow hair. Photographers have captured pictures in which some of the hair is green! It is algae that are growing in and on the hair. Since the hair is actually colorless, the green shows through.

Ice bears are omnivorous. However, on ice, there are no plants and berries to eat, so their diet is meat. Humans are meat. So each naturalist carries a flare gun and a rifle. The first line of defense is to stay away from bears! Expedition staff members go in Zodiacs around the landing sites to search the land and sea for bears within a mile of the ground party. Second line of defense, if a surprise bear makes an appearance, is to fire the flare gun close to the bear. This creates an explosive sound, extremely bright light (like fireworks), and a smoke barrier between you and the bear. Only if those efforts fail would a rifle be used. Our naturalist was well-trained in her firearms but told us she had never had to use the rifle.

S   Scat

Can you figure out what the ice bear eats by looking at its poop?

How about for the reindeer pellets? Could you get some clues about the reindeer's diet from looking at WHERE it poops?

Big Word: Sulfur Dioxide!

Additional Information:

Fossil Fuels, such as coal and petroleum, contain sulfur compounds. When they are burned, sulfur is released into the atmosphere as sulfur dioxide, which interacts with water and creates "acid rain". Other sources of sulfur dioxide include the burning of wool, hair, rubber from vehicle tires, and foam rubber from mattresses, seat cushions, and carpet pads. Sulfur dioxide is a major air pollutant and has significant impacts upon human health and influences habitat for plant and animal life as well. The United States has dramatically reduced sulfur dioxide emissions by 94% since 1970.

Seasickness: It is common wisdom that this annoyance is caused by the rocking motion of a vehicle, such as a car or boat. Why should motion cause nausea? When we run or dance or ski we don't get seasick. We are giving our intestinal tract much more intense and dramatic shocks than when we are riding in a vehicle! And generally we have no abdominal side effects from those activities. The real cause of the nausea is in the brain, which is getting conflicting signals: your eyes show a world that is still while your body (especially the equilibrium sensors in the ears) feels a world that is moving. This conflict causes the brain to send a "whole body alarm" to stop all activities, including one of the most complex of the bodily functions, digestion.

T   Tundra   Tusk

The tundra looks pretty dead from a standing position, but when you move down close, you can see a miniature forest of plants.

If you had a tooth this big, how would you use it?

Big Word: Thigomatic!

Additional Information:

Walruses are thigomatic; they are "touchy, feely", extremely social animals. They lay together on the beach to conserve heat, to mate, and to congregate socially. The piles are segregated by "role". Young males haul-out together. Females and calves haul out separately from the adult males. Pregnant females will haul-out by themselves for giving birth, but quickly rejoin the others in the herd. Walrus cows nurse the young calf for 2 to 3 years. Then female young join the herd, while male young go off to join the other young males.

Walrus tusks are up to 3 feet long. The main role of tusks is social— to show dominance and for mating displays. Both males and females have tusks. The length of the tusks can be an indicator of age until the walrus is an adult. Some anecdotal reports mention the use of tusks as weapons, as hooks to help the walrus haul-out, and as a "sled" to help glide along the sea bottom for feeding.

Walrus mostly eat clams, sucking the clams right out of their shells and leaving the shells on the bottom of the ocean. They dive up 300 feet and can stay under for up to 10 minutes. Walruses came under protection in Svalbard in 1952. At that time there were only a few hundred individuals left. Now there are probably 2-3,000 animals.

U Ursus Maritimus

What do you think "Ursus" means? Tell a story
What do you think "Maritimus" about what is
means? How many ursii maritimii happening
are part of this situation? here!

Big Word: Ursus Maritimus!

Additional Information:

Ursus Maritimus actually means "sea bear". We call them "polar bears" because they live only in the North Pole regions of the world. We also call them "ice bears" because we usually see them on ice flows. They need the sea ice for survival. With their main food being seals, they need to be where the seals are, which is in the ocean. Seals make breathing holes in the ice. Ice bears sit patiently at a hole, waiting for a seal to come up. The seals are pretty smart. They will come up to a hole and blow some bubbles up. If a big white paw comes out to "grab" the bubbles, they know it is not safe to breathe there and will go somewhere else. However, a clever sea bear knows not to swipe over the hole and will wait for the seal to surface. Then the bear's two-inch long claws can grab the seal and pull it out of the hole for a patiently-awaited meal.

When sea ice melts, an ice bear may be stranded on land. Then there is a real possibility that it will starve to death before the next sea ice season arrives three or four months later. Global warming trends are hurting the ice bear population in all the polar regions by lengthening the sea-ice free summer in more southerly regions and decreasing the area of more northerly sea ice that remains during the summer months.

V Vikings

Is this for real? Are there any Vikings exploring the seas in Zodiacs? Did Vikings in ancient times really wear hats with horns?

This helmet is the only known complete Viking helmet. Many scraps of helmets have been found, but NONE of them have any horns!

Big Word: Vikings!

Additional Information:

The pictured helmet is made of iron. Viking helmet parts have only been found in three sites. This one was found in Gjermundbu, Norway in 1943. It is dated to 970 AD. Helmet pieces have also been found in Denmark and Gotland Island, Sweden. Icelandic and Scandinavian stories and picture stones show warriors with helmets made with leather strips and iron pieces. Some believe that the helmets were passed from father to son over the generations and so were worn out to the point of being turned into scrap for other uses. The horned and winged helmets associated with the Vikings in popular mythology were the invention of Romantics in the 1800s.

W Whale Walrus

Hump and Dorsal Fin   Blow Hole

Hump-Back Whale

Notice the barnacles attached to the tips of the whale's tail! Think of all the places they have traveled!

Walruses "haul-out" onto the rocks to lie in the sun to warm up. Arctic water is COLD!

Big Words: Baleen! Krill!

Additional Information:

Baleen is a filtering system inside the whale's mouth. Think of putting a flexible comb around your teeth. Open your mouth and swim through water that has food in it. Then close your teeth and spit the water out through the comb. You'll be left with a mouthful of delicious krill! That is how the hump-back whale eats! The baleen comb can be up to 11 feet long. It is made of keratin, similar to human fingernails.

Krill are small crustaceans. Shrimp are in the krill group of sea creatures. Krill are swarming animals. In the polar arctic, the krill are small and the swarm might be over 10,000 krill in a cubic meter. That is easily the size of the mouth of the whale, so the whale might get 10,000 krill in one gulp!

Global warming is changing the distribution of krill because each water temperature area has a different species. Krill are an indicator species of sea animal. Krill eat algae that form under sea ice. As sea ice disappears, so do krill.

X   Saxifrage

Purple Saxifrage

Tufted Saxifrage

Big Word: Saxifrage!

Additional Information:

Saxifrage is one of the most prolific of tundra flowers. This is a good thing because it is a staple in the diet of reindeer (remember the picture of the reindeer scat pellets on the "R" page? Where were they?) and arctic hares. Both the flowers and the leaves are edible. They grow to about 2 inches in height. So the animal has to nibble very close to the ground. It spreads along the surface with creeping, woody stems. Each stalk has a single flower that is smaller than your fingernail.

The name saxifrage derives from the Latin for "rock breaker". The root of the saxifrage is a stout taproot that pushes down into cracks in the rocks, breaking them up over time. The taproot stores up carbohydrates to sustain the plant through the extreme cold and windy conditions of the tundra. The flower looks like a bell as it starts to bloom and then unfolds into a five-petal star. If any of the flowers survive the foraging of reindeer and hares, a small fruit filled with seeds will develop within a few weeks. The growing season is very short, sometimes just two or three months from last snow to first snow. The leaves and stems of saxifrage, like many tundra plants, are furry. This helps insulate the veins of the plant from the cold. Also, growing in a mound, instead of spreading out, helps keep the warmth close by the roots.

Y   Yamaha   Yearling

way to get around in the winter when the sea is frozen. There aren't any roads except right in Longyearbyen.

Every home in Svalbard has a snowmobile and various sleds because it is the only

This walrus calf is a yearling and two tusks have started growing out of its mouth! Why don't humans have tusks growing out of their mouths?

Big Word: Snowmobile!

Additional Information:

There are few roads in Svalbard—just 27 miles in an area about the size of the Mojave Desert in the US, or the state of West Virginia or the countries of Latvia or Lithuania. So travel within the archipelago is restricted to snowmobiles, dog sleds, airplanes, and boats when there is no ice. The airport, at Longyearbyen served 125,000 passengers in 2011, an average of just 350 passengers a day. About 2 million pounds of cargo also came in and out of the airport.

Snowmobiles are made by six major manufacturers: Arctic Cat, Yamaha, Polaris, and Ski-Doo (Bombardier Recreational Products), Snowhawk (Boivin) and Alpina. Prices range from $7,500 to over $15,000. In contrast, sled dogs cost $500 to $1,000 each. In both cases, you also have costs of upkeep and storage, feed or fuel, repairs or injury mitigation, and equipment such as sleds, skis, harnesses, etc. Sled dogs are not pets, just as snowmobiles are not children's toys. In Longyearbyen, dogs must live in one of the four city approved kennels and can be taken out for sled training year round, but not for just playing in the yard!

Z   Zodiac

A crane on the ship lifts the zodiac off the pile, and sets it down on the sea. If you had a crane this size, what would you do with it?

The zodiac uses gasoline for fuel. Even though it is a very small motor, there is exhaust! How far do you think a zodiac could travel on one tank of fuel?

Why do you think it is so awkward to get in and out of a zodiac?

Big Word: Zodiac!

Additional Information:

A zodiac is an inflatable boat. It is used for short excursions and rescues. They are great for landing on beaches and are also used in sporting events and recreation, such as white water rafting.

The Naturalists onboard took the Zodiacs out before every exploration to scout for bears in the 2 or 3 miles surrounding the planned landing. A bear sighting, on land or ice, or in the water is cause for cancelling the planned excursion. They are very hungry in the beginning of the summer, and therefore, very dangerous

Climate Change in the Arctic

Like in all parts of the earth experiencing the global warming phenomenon that is happening, the climate changes in the Arctic are affecting animals and plants. The Arctic Council is an intergovernmental forum of Arctic countries; they have established a working group to study these issues. Countries include Canada, the Kingdom of Denmark, Finland, Iceland, Norway, the Russian Federation, Sweden, and the United States. Working groups make comprehensive, cutting-edge environmental, ecological, and social assessments. https://arctic-council.org

As arctic temperatures rise and permafrost melts, microbes work on the newly thawed soil which converts to carbon dioxide and methane and other greenhouse gases such as sulfur dioxide. In addition, new species of plants begin to move in, replacing the traditional food source of reindeer.

As the graph on the "C" pages shows, the decline in the extent of arctic ice has been dramatic over the last 4 decades. Many predictions can be made, based on these observations. For example, the Arctic Ocean may become sea-ice free in the summer as soon as 2050. Polar bears and walruses are ice- dependent animals who are put at risk as ice melts. Starvation and reduced births endanger these populations.

An oddity, you may think, is that when ice melts, the sea temperature rises! Really? Shouldn't it go down because the ice was cold before it melted? NO! Ice reflects sunlight heat, while water absorbs sunlight heat. Higher water temperatures affect the food and living conditions of marine mammals. White beak dolphins (see "D" pages) lose habitat. Walruses and shore birds' food supplies change.

On the land, warmer temperatures increase parasites and diseases and damage food sources for reindeer and arctic foxes. In addition, red foxes are moving north to compete with the arctic foxes for denning sites and food supplies.

In the atmosphere, changes due to permafrost melting cause releases of green-house gasses as soil microbes convert carbon into carbon dioxide and methane.

Using the Internet to further explore many of the topics in this book is easy! Public libraries have internet computers to explore some of these sites.

American Museum of Natural History: Ology Home https://www.amnh.org/explore/ology/search/(keyword)/arctic puzzles, coloring pages, interactive story puzzles, quizzes.

NASA Climate Kids https://climatekids.nasa.gov/kids-guide-to-climate-change/ excellent, easy to understand guide from NASA, with lots of extra links to activities and other resources.

Jean-Michel Cousteau Ocean Adventures https://www.pbs.org/kqed/oceanadventures/educators/arctic/ Lesson Plan: Arctic Animals and a Changing Climate, grades 5-8.

Center for Climate and Energy Solutions https://www.c2es.org/content/climate-basics-for-kids/ great explanations for kids who are tweens, not really for little ones.

What You Can Do:

Check boats for barnacles or other marine life attached to the underside and clean them off so you don't transport them to a new environment where they don't belong. 💙 Shake out jackets and back packs and clean off your shoes or boots when you leave a space, so you aren't taking plants, seeds, and bacteria to new environs where they don't belong. 💙 Be quiet around wild life. Respect their need for feeling safe. Don't yell or try to scare them, unless you are doing it to protect your own life. 💙 Don't leave stuff behind! Take your trash out with you! "Stuff" also means your actions! Don't rearrange the landscape or tear down natural formations. 💙 Don't put yourself in danger by going into a space that is home for a predator animal.💙 Make sure your carpets and pads get recycled by the carpet industry rather than sent to a landfill where they may release sulfur dioxide into the environment. Likewise with old tires. Let the tire dealer dispose of them properly!

💙 Figure out what might be a good indicator species for your neighborhood. Go out and count.

As you read this book together, if you think of more ideas of *What You Can Do*, send them to jgasper2@iCloud.com. Maybe they'll show up in the next edition of this book!

Photography Acknowledgements—All others by Joseph and Juli-Ann Gasper

A   Arctic Circle. Wikipedia commons license.

     Svalbard Archipelago. Wikipedia commons license.

F   Arctic fox. Rachel Hopper, by permission.

     Arctic fox, winter coat. Public domain.

     Arctic fox, curled. Creative commons license.

G   Guillemot eggs. Henry Seebohm. Public domain.

J   Parasitic Jaeger. Rachel Hopper, by permission.

     Fox. Pierre Vernay. ARKive.com license.

K   Kayak. Suzanne M. Edmonds, MD, by permission

O   Oyster. Wikipedia commons license.

     Mussel. Public domain.

     Limpet. Wikipedia commons license.

     Crab. Sarah Minks. Hardy, UAF.

     Oystercatcher. Rachel Hopper, by permission.

P   Rock Ptarmigan. Rachel Hopper, by permission.

     Svalbard Poppy. Ann Inoye, by permission

Q   Ivory Gull, Public domain.

     Oystercatcher. Rachel Hopper, by permission.

     Barnacle Goose. Rachel Hopper, by permission.

     Puffin. Mark Prior, mrp@mrp.net, by permission.

     Arctic Tern. Jillien Cruse, by permission.

     Northern Fulmar. Mark Prior, mrp@mrp.net, by permission.

R   Reindeer, Coleman C. Eaton, III, by permission.

S   Bear Scat. Ann Inoye, by permission.

V   Viking helmet. John Erling Blad, creative commons license.

VOCABULARY LIST

(shows letter pages where found)

Crustacean	B	G	W	Forage		X
Database			H	Fossil Fuel		S
Decontamination			D	Fox		F
Dehydration			L	Fulmar		Q
Denmark			V	Fungus/Fungi		L
Digestion			S	Furry		X
Disembarkation			D	Geolocator		Q
Dog		B	Y	Geology		N
Dog Sled			Y	Glacier		G'
Dolphin		D	K	Global Warming	I M U	W
Dorsal fin			W	Gotland Island		V
Ecosystem		N	O	Greenhouse Gases	M N	S
Education		A	N	Grouse		P
Eggs			J	Guillemot	F	G
Elliptical			M	Gull	J	Q
Engineer			M	Hair	R	S
Environmental Studies			N	Haul-out	H T	W
Equator			A	Helmet		V
Equilibrium			S	Herd		T
Explorer		B	E	Horticulture		N
Feather			G	Hump-Back Whale	H	W
Fingernail		X	W	Ice Bear	B J R S	U
Fingerprint			H	Ice Cap	A C	I

58

Onomatopoeic	K	Ptarmigan		P
Orbit	M	Puffin		Q
Organic	M	Reindeer	O R S	X
Oslo	A	Rifle		R
Oyster	O	Ring Seal		F
Oystercatcher	O Q	Role		T
Ozone	N	Rubber		S
Parasitic Jaeger	J	Russian		A
Particulate	S	Sanitize		D·
Partridge	P	Saxifrage	E O	X
Peat Moss	M	Scandinavia		V
Pellet	S	Scat		S
Penguin	Q	Scavenger		Q
Permafrost	D	Scientific Research		A
Petroleum	S	Sea Bear		U
Philippine	A	Sea Ice	A B U	W
Photosynthesis	L	Sea Kayak		K
Pinniped	C	Seagull		K
Plumage	Q	Seals	B C	U
Polar Bear	B U	Seasick		S
Pollutant	N O S	Seaward		C
Poop	S	Shock Absorber		G
Poppy	P	Shrimp		W

Way Up North Where the Kittiwakes Play:
An A to Z Alphabet Book for Child and Adult
by Juli-Ann Gasper
MainSpring Books

book review by Kat Kennedy

"Svalbard is home to a distinct population of reindeer; small, with short legs, snubbed muzzles, and chubby bodies."

Inspired by a National Geographic Society-sponsored expedition to the Svalbard Archipelago, Norway, this alphabet book offers fascinating facts and photos about this far north land and its inhabitants. Both animals and concepts are presented using the alphabet. Each letter and the accompanying photos reflect knowledge gained by the author on the expedition as well as her extensive research. Questions are presented for further comprehension of each segment. The book offers parents ideas on how to use each alphabet page and its additional information page. There is also a vocabulary list at the end, which works as an index for finding information in the book. The author also includes links to websites for further study.

Gasper has created a wonderful source for children and adults in this alphabet book. The information can be tailored to even the youngest readers as the photos will foster their imaginations. Each alphabet page and its accompanying information page is filled with captivating details. The author includes notes on how to use the book and questions in each section to further understanding. There is information on the Arctic Circle and the specific species that live there. For example, for the alphabet letter F, there are pictures of the arctic fox with leading questions such as: "What season is this? What makes you think that?" There is also extended information about the arctic fox: "It has several adaptations for cold survival, including very deep thick fur, high supply of body fat, a stubby body to reduce surface area exposed to cold, and short thick ears. Also, the fox has a countercurrent heat exchange system in the blood vessels of its legs to keep its body heat concentrated in the body." The beauty of the way information is presented in the book is its easy adaptation for any age. It could also be used in science classes in elementary school or even higher grades.

This is not only a book for children, however. There is plenty of information to keep adults interested in the book. For example, under the letter J the parasitic jaeger is introduced. On the information page, readers are given the "big word" kleptoparasite, meaning "it pirates food from other animals!" Examples are given of other arctic animals who steal from others. "Gulls that can't dive will steal the fish that diving birds bring to the surface." The letter K gives us insight into the book's title. "Kittiwake is a seagull." Following is information on the kittiwake, including "If you say it like 'kitti waaaaaake', it sounds like their call." Information like this is sure to instill interest in young readers and parents alike. The book is filled with interesting details such as this on every page, making it a wonderful resource book for the home library.

This is an intellectually stimulating book that will delight both children and parents. It is also a wonderful book for adults as it is filled with captivating information. The photos are plentiful and include a wide variety of settings and animals. The book captures the imagination and offers a great deal of information to keep readers returning to its pages.

RECOMMENDED by the US Review

HOLLYWOOD
Book Reviews

Way Up North Where the Kittiwakes Play: An A to Z Alphabet Book for Child and Adult by Juli-Ann Gasper transports readers into the mesmerizing realm of the Arctic, beckoning individuals of younger ages, as well as adults sharing in this reading experience, to delve into its marvels through the alphabetical framework. The adventure begins right from the start, immersing readers in a captivating exploration filled with wonder and learning. Gasper combines beautiful pictures with engaging stories, creating a perfect blend which educates and entertains at the same time.

Gasper has crafted an exceptional educational resource in this alphabet book, suitable for both children and adults alike. The vivid photographs stimulate the imaginations of young readers while providing enriching content for adult readers. For example, younger readers will be mesmerized by the colorful illustrations and simple descriptions accompanying each letter. For example, let's look at the letter "P" for polar bears. In this part, Gasper draws a beautiful picture of the polar bear in its natural habitat with icy water and snowy mountains. The words that go along with it give us some interesting facts about polar bears, making it the perfect introduction to Arctic wildlife.

On the other hand, adult readers will appreciate the depth of information provided on each page. Her writing strikes a perfect balance between educational content and lyrical prose, ensuring that readers can absorb new information while enjoying the narrative.

By highlighting big words and adding additional information, the book not only explains concepts but also offers scientific explanations behind them.

One of the main plot points of the book is conservation and environmental awareness. Gasper shows how important it is to take care of the Arctic and the animals living there, especially with things like climate change and people harming the environment. This exploration goes further by providing readers with insightful glimpses into the customs and heritage of the indigenous communities intimately connected with the Arctic's terrain and seas. In addition to enriching understanding, each segment is accompanied by stimulating questions, providing valuable suggestions on how to interact with the content found on every alphabet page and its associated information page. The help of a vocabulary at the end leads to navigating through the book effortlessly, accompanied by links to further online resources for those interested in delving deeper into the subject matter.

The illustrations used in the book are breathtaking. In the examination of the letter "I" for ice caps, Gasper offers readers valuable insights into the formation and importance of these frozen features in the Arctic. By illustrating how ice caps function as vital habitats for diverse Arctic wildlife, such as seals and polar bears, Gasper enhances readers' comprehension of the Arctic ecosystem and the intricate relationships between its residents. Another example could be the letter "F" for Arctic fox; Gasper not only introduces children to this fascinating animal but also provides adults with exciting insights into its unique adaptations for survival in the harsh Arctic environment. Gasper's inclusion of details such as the fox's thick fur and stubby body enhances the educational value of the book for readers of all ages.

Way Up North Where the Kittiwakes Play is engaging and enlightening. Its abundance of knowledge and visually striking illustrations make it a worthwhile asset to any home library collection.

Pacific Book Review

helping authors succeed!

Rarely do I read a book with such an impact in a children's genre with vocabulary, information, imagery and education, as *Way Up North Where the Kittiwakes Play* by author and photographer Juli-Ann Gasper. During a trip to Norway's Svalbard Archipelago in July of 2012 aboard the *National Geographic Explorer,* Juli-Ann and her husband Joseph have compiled a cornucopia of information and images which adorn this book's pages, along with a photo-background on each page of ice-laden ocean photographs - which actually made me subconsciously feel the coldness and frost of the region.

In a multi-informational format, the author presents fact after fact about life in the Arctic Circle, combining biology details about the wild animals and fauna, geography about the location on Earth, and interesting ecological details about the impact civilization has had upon the region. All of this is done with a stylish written-narration of confidence and intelligence, drilling down on the facts to the basic terms - for the entire reading audience's comprehension and benefit. But it doesn't mean she skimps over "large words" but rather defines the meanings and shows examples of how these concepts are important. This is a seasoned craft of teaching which is explementary.

For example, what is a Saxifrage? She explains it being one of the most prolific of tundra flowers, and the staple in the diet of reindeer. The definition of the word "Kleptoparasite" can be deduced by "Klepto" from stealing and "parasite" from eating from another source, hence the bird called the *Parasitic Joeger* is a kleptoparasite because it steals other animal's meals. Also, one of my favorites, a *Guillemot;* a sea bird which lives on vertical cliffs along the shore. The shape of their eggs is such that they don't roll, but if nudged, they spin – thus not falling off the cliff. Amazing.

The format of the book is extensively cross-referenced with both an index at the end and an alphabetical listing of many of the specific words used in a glossary. The author asks many questions throughout the book which will encourage a dialogue between adults and children paging the book together. The amount of information contained will embellish the understanding of both children and adults alike.

Way Up North Where the Kittiwakes Play, An A to Z Alphabet Book for Child and Adult is a book not to be missed, and a must-read to be enjoyed by all members of families living below the Arctic Circle. It will reveal amazing facts to all readers, regardless of their age, but most certainly appeal to the limitless curiosity of youngsters.

READERS' CHOICE BOOK AWARDS

An arctic-themed alphabet picture book by Juli-Ann Gasper.

Way Up North Where the Kittiwakes Play: An A-to-Z Alphabet Book for Child and Adult is a delightful arctic-themed picture book by Juli-Ann Gasper, with stunning photography by Joseph and Juli-Ann Gasper. The stories and pictures in the book are taken from an expedition to Svalbard Archipelago, Norway, in July 2012. From A (Arctic Circle) to Z (Zodiac), this alphabet book provides fascinating insights into the landscape, inhabitants and culture of the Archipelago. Did you know that cats are banned on Svalbard because they attract bears? Or that Arctic Tern's have the longest migration of any known animal? Or that Walruses are thigomatic, in other words, extremely tactile and sociable animals? Each page provides a description of the subject associated with the letter of the alphabet, alongside fascinating facts and figures. Each letter of the alphabet is accompanied by beautiful, colorful photography, which showcase the beauty and diversity of the Arctic Circle.

The book uses the structure of the alphabet to introduce readers to the unique wildlife, breathtaking landscapes and rich cultural heritage of this remote region. With lots of additional information at the end of this book, this alphabet book is a great resource for budding naturalists, explorers and anyone interested in nature or the environment. Both educational and entertaining, the book will appeal to young and adult readers. The content is accessible for young audiences, and it will help to build knowledge and vocabulary. The book also provides more depth information for older readers.

Star rating: 4 Stars

Summary: A captivating A to Z picture book, exploring the icy realms of the Arctic Circle.

Review completed by Readers' Choice Book Awards

CERTIFICATE *of* EXCELLENCE *in* CHILDREN'S LITERATURE

Winner
Children's Non-Fiction

Way Up North Where the Kittiwakes Play

Juli-Ann Gasper

Linda F. Radke, Contest Director

STORY MONSTERS

Purple Dragonfly Book Award
SM WINNER

2024

CERTIFICATE *of* EXCELLENCE *in* CHILDREN'S LITERATURE

Green Books/Environmental

Way Up North Where the Kittiwakes Play

Juli-Ann Gasper

Linda F. Radke, Contest Director

STORY MONSTERS

Purple Dragonfly Book Award
SM WINNER

2024

www.ingramcontent.com/pod-product-compliance
Lightning Source LLC
Chambersburg PA
CBHW040848120626
46547CB00001B/74